Integrated Interactions at International Junior Science Olympiad

DR P.K. JOSHI

गरुड

Published by
Garuda Prakashan Private Limited
Gurugram, Bharat

www.garudalife.in

First published in India 2025

ISBN: 979-8-88575-252-7

Cover Design: Bipro Kumar Sen

Printed in India

Contents

Preface

International Junior Science Olympiad (IJSO) is a competition in science for students who are fifteen years old. IJSO aims to promote interest in science among school students, exposing them to problem solving, critical thinking and experimentation. Over the years IJSO aims to positively impact science and mathematics education at school level. The short articles here are based on the questions prepared for the selection of IJSO students who represented India at the International competition on several occasions. I have included some of the topics in which I contributed the major portions. It also contains problems discussed with my colleagues and occasional inputs from the team members of JSO who are acknowledged at appropriate places. The other problems discussed by other team members will be discussed in a separate compilation.

The articles go beyond the question-and-solution format by describing the thought process behind the questions, the detailed explanation of the solution and many other related descriptions. Hence, some of the assumptions in the topics deviate from reality but are not un-realistic and the further analysis is based on those simplistic assumptions for school children.

The target audience is the class VII and X students and their teachers with the purpose of introducing them to challenging topics of nature which involve several different concepts of science and mathematics, often at the same time. My intention had been to expose the students to concept of integration of different sciences, a key feature of IJSO.

During my leadership, several interesting problems were designed for the selection process of the Jr. Science Olympiad students who represented India at the international event every year. During this period Indian teams did exceptionally well at the IJSO. In 2007, the first year, three students participated and all three received silver medals. In 2008, a full team of 6 students represented India and the team received 5 silvers and one bronze medal. This later turned out to be the only bronze medal India ever received in IJSO under my leadership. The performance of the team peaked in its last 5 years. Out of 30 participants, India received 25 gold and 5 silver medals. In these years the world ranking of India was 1st, 2nd, 1st, 3rd and 2nd, respectively. These glorious years of IJSO were also marked with Experimental Gold Medal and Experimental Silver Medal. I also had my share of personal achievement when I was selected as the vice-President of IJSO from 2012-2015 and then as President of IJSO from 2015-2022.

Though the chapters in this book are based on questions of JSO, they aim to strengthen the learning process of the students. Hence, the articles are intended to make interesting observations in science which can help students enhance their knowledge base, which will help them in tackling not only competitive exams of various kinds but also prepare them for real-life situations in their future career.

PK Joshi

ACKNOWLEDGEMENTS

I would like to acknowledge the contribution of many of the Junior Science Olympiad team members who have contributed in terms of vetting out the equations or clarifying/simplifying concepts at different stages of the problems. Hence their contributions change from topic to topic and in different capacities. The major contributions have been acknowledged in respective examples.

But in general I would like to thank all the members of the JSO team who were involved at various stages for their valuable inputs to different examples. Many of these issues were discussed in the group and hence it is not possible to identify the exact contributions received from the team members. And hence in general I thank each one of them irrespective of their contributions. I fondly acknowledge contributions of Dr Ritesh Khunyakari, Mr. Vinayak Katdare, Mr. Pradeep Dasgupta, Mr. Zohar Ataree, Prof. Pradeep Burma, Dr Sachin Purohit, Dr Shashibhal Pandey, Dr A.K. Rajarajan.

I would also like to express special thanks to the JSO staff members, Jueeli Lad, Bhagyashree Chemate, Shuvecha Mukherjee and Prakash Nawale for their support in data collection, editing and formatting.

Finally, I would like to thank Prof Deepak Mathur (Retired, Department of Nuclear and Atomic Physics, TIFR and currently at Manipal University) who has been a constant source of inspiration and encouragement to Junior Science Olympiad and personally to me. He was the Chairperson of the Academic Committee of IJSO-2013, hosted by India. He has also vetted all the material put out in public domain, related to Jr. Science Olympiad, including this piece of work.

Framing the questions was a group activity among the members of IJSO. All the experts (teachers from schools, colleges, research institutes etc.) discussed each question jointly. The integration process was a boon to all the members because teachers with physics background started seeing potential of biology concepts in a given question and the other way around. At the end of every session the teachers were seen discussing these new aspects with great enthusiasm.

Due to this team event of forming theory and experimental questions, it was often really difficult to gauge the quantum of contribution from any particular individual. The positive part of this group activity was that the questions were formed by an active body of academicians who gave their best. And, individual ownership was almost not there. The other positive outcome was that there was a very high collaborative attitude among the team members leading to developing a team of satisfied members.

A social outcome of these meetings was that the team members gelled very well to work in subgroups which were formed on the basis of the question being discussed. Thus, new subgroups were formed for every question, which had members from across all the three branches of science. The members experienced great sense of satisfaction at the end of the meetings of having a great academic achievement.

PK Joshi

01

Introduction

International Junior Science Olympiad, IJSO, has a unique feature of integration of sciences. This is very essential in studying and teaching sciences as it helps in developing a holistic approach of learning and understanding science and mathematics topics. This also makes science very interesting and attractive by bringing the students closer to nature. Some simple examples would clearly explain this.

At the selection and training camps of Junior Science Olympiad many questions were raised by the students. These questions and subsequent interactions with the author are discussed in the subsequent sections. Many articles go beyond physics, chemistry and biology to include mathematical concepts and several other general science topics.

In a general (qualitative) survey conducted during the programmes with both students and teachers, a question was asked to give the most common application of convex lens in our day-to-day life. The standard answers received were microscope, magnifying lens, and to some extent telescope. Upon probing further, insisting that these are not 'the most common' applications, the next popular answer was spectacles. However, upon pointing out that since half of the participants were not wearing spectacles, how can it be 'most common' example? It took further prodding to get the answer as "eyes". This is because students and teachers have trained themselves to put lens in "physics" category/ compartment whereas "eyes" come under biology, and certainly there is no bridge between the two.

Integration is the need of the hour and the pathway to further revolutionary research and advancement in sciences. Many of the path-breaking discoveries are now more and more integrated in nature. This book deals with some very interesting topics covered in JSO, but a few of them deal with the concept of integration of sciences.

Let us look at some interesting examples along these lines based on the questions discussed at the Orientation-cum-selection camp of IJSO held between 2008 and 2018.

In all, the examples/assumptions are for class VII to X students to understand the concepts behind the science phenomena. The assumptions can be modified to be more realistic if a student is interested to probe deeper into the topics to get a much clearer picture on the phenomena discussed.

Many of the assumptions may even be deviating from reality but do not fall under "impossible" or "unscientific" category. The aim of these examples is to give students a "feel" for the quantum of the issues involved.

02

Rise of water in a Neem tree and its velocity

While having lunch at the HBCSE canteen (in the outer area) the participants were joking about how fast one can eat and drink to go back to the session. In the middle of these light moments, one of the participants asked me, “Sir, how fast does this tree (pointing towards the adjacent Neem tree) ‘drink’ water?”

After a pause I answered, “Well, we can either consult the Internet or we can make our own estimates, which may not match with the exact number but can give us insight into many more concepts involved.”

Figure 1. Neem tree

Moving into the classroom after the lunch break, we started discussing the Neem tree.

A Neem tree (*Azadirachta Indica*, Fig. 1), like any other tree, gives out oxygen during the process of photosynthesis. It has been experimentally established that during the process of photosynthesis, oxygen splits out from water and not from carbon dioxide. However, this important feature of photosynthesis is often ignored while teaching about photosynthesis. The release of oxygen from water was experimentally confirmed by use of radioactive oxygen in water.[1] The radioactive oxygen was not observed in glucose or water, whereas the oxygen gas released was radioactive.

Figure 2. A typical neem leaf

It is known that this tree gives out 260 lb of O_2 per year. Therefore it can be easily deduced that they give out 13.5 g of O_2 molecules per hour. This number is not a sacred number but a simple assumption close to the estimated number available in literature.[2] In order to generate the oxygen, tree uses blue (wavelength 475 nm) or red (wavelength 700 nm) light for the process of photosyntehsis. It is also established that for every 8 photons of incident radtiation (on an average) one O_2 molecule is released in atmosphere.[3]

For simplicity it is assumed that only blue and red wavelengths contribute to the process of photosynthesis.[4] However, the contribution from other wavelengths is non-zero.

Let us assume that the foliage of leaves is making a hemisphere (half sphere of 10 m diameter)[5] standing on a trunk which is 50 cm in diameter and average area of a neem leaf is 11 cm^2. This can be seen in Figure 2, which shows a typical neem leaf placed on a graph paper and its area calculated. Area of 6 random leaves was calculated and the areas obtained were 10.8, 13.2, 11.2, 10.8, 9.8 and 7.9 cm^2, which gave an average value of 10.6 (±1.7) cm^2. Students who are not aware of the concept of (±1.7) can ignore the number till they understand its significance.

As per the The Forestry OutReach Site (http://dendro.cnre.vt.edu/forsite/tait2.htm), "The xylem is composed of dead, thick cells that act as pipes for transporting water and nutrients up the tree. The active portion of the xylem is known as the sapwood. The sapwood is found near the outside of the tree. Older xylem is known as heartwood and is found in the middle of the trunk."

The authors[6] have measured the diameter of the tree at breast height (dbh) (4.5 ft = 1.37m) from the base of the tree and also measured the width of the sapwood. As per the measurement, the order of magnitude of the ratio of xylem (functional) width to the dbh is around 1:10.

It is assumed that 10% of the tree trunk cross section area corresponds to xylem tissues. These xylem tissues are responsible for transporting water up the trunk from the roots. [7]

Photons of different colours of light (wavelengths) fall on the canopy. They are absorbed by the leaves for process of photosynthesis. Using the data available in graph (Fig. 3) one can calculate several aspects. From the graph, for blue light the value of wavelength $\lambda = 475$ nm. Thus,

$E = hc/\lambda = 6.63\times 10^{-34} \times 3 \times 10^{8} / (475 \times 10^{-9}) = 4.19 \times 10^{-19}$ J and the number of photons is 4 μmoles.

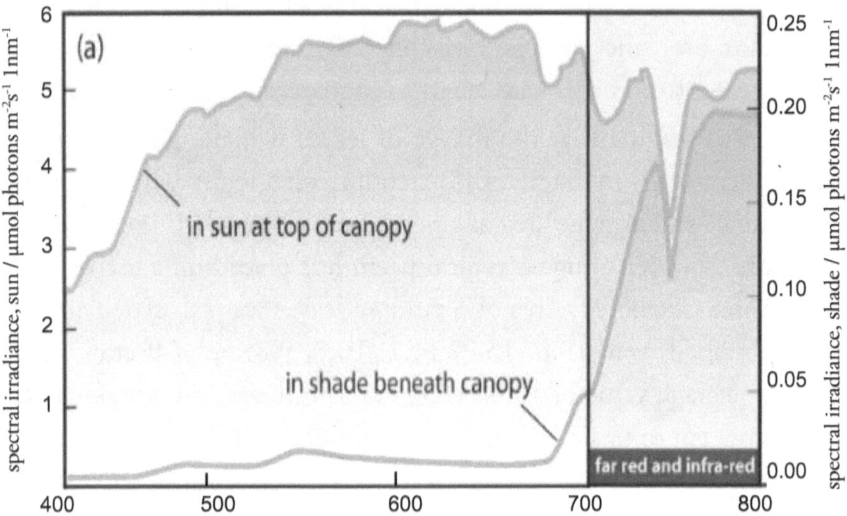

Figure 3. Graph of intensity of radiation as a function of wavelength of radiation[8, 9]

Thus, number of photons = $4 \times 10^{-6} \times 6.023 \times 10^{23} = 2.41 \times 10^{18}$.

The energy delivered is 1.0088 J/s. Here 1 mole of irradiance is 6.023×10^{23} photons of that wavelength.

Similarly, for the red light (700 nm) photon has 2.84×10^{-19} J energy. Number of photons is 3.01×10^{18} and so total energy delivered is 0.8557 J/s.

Average number of photons incident on the surface of earth, every second, as available in the literature, is 2.71×10^{18} per m^2. (This number is the average of 2.41 and 3.02 ($\times 10^{18}$), calculated earlier.)

The total amount of oxygen released per hour is 13.5 g, i.e., $(13.5/32) \times 6.023 \times 10^{23}$ number of O_2 molecules per hour = 2.5×10^{23} molecules per hour = 7.05×10^{19} per second.

Study [3] suggests that 8 photons are required to release one molecule of O_2 gas. Due to the efficiency of conversion, to generate this oxygen 8 times more photons are required, i.e., 5.6×10^{20} photons.

Hence the area which can match these numbers is $A \times 2.71 \times 10^{18}$ per m^2.

This area, to capture required photons, is therefore calculated as $5.64 \times 10^{20}/2.71 \times 10^{18} = 208\ m^2$. This area is provided by the foliage of the neem leaves which have average area of 11 cm^2. This implies that the foliage has 0.2 million leaves.

The foliage is 5 m radius hemisphere. The total volume of leaves is therfore 262 m^3. This implies that the tree has 720 (approx.) leaves per m^3.

Process of Photosynthesis is defined by the equation

$$6CO_2 + 12H_2O \rightarrow C_6H_{12}O_6 + 6O_2 + 6H_2O.$$

Since the total oxygen released is 7.1×10^{19} molecules per second, 2 times more H_2O is required.

Water molecules therefore required = 1.4×10^{20}. This amounts to 421 mg of water per second. But only 1% of water which is supplied by roots is used for photosynthesis and thus amount of water supplied by roots is which is equal to 0.421 ml of water, per second.

As per the publication[10] the radius of the neem tree increases by around 1 cm per year. So, let us assume the radius of the tree trunk to be 25 cm giving an area of $\pi 25^2$. Assuming that 10% of area has the (functional) xylem cells, total cross sectional area of xylem would be 196 cm^2. The height of the water column rising per second would be 21 μ.

Hence the velocity of water is around 21 μ/s.

Here, we must remember that the quantity of sunlight falling on the tree, the number of leaves on a tree, the area of the leaf etc. can vary drastically from case to case. Thus the velocity of the water rising also can vary depending on all these parameters; however, the velocity will be in the range of few microns per second. We must also remember that these numbers are not the ultimate. For a research project, more precise measurements would be needed.

I would like to thank Dr Ritesh Khunyakari for his valuable inputs for this example.

03

Concept of infinite series

One day I had just started discussing the different kinds of forces in nature and was discussing the gravitational force which exists between any two objects.

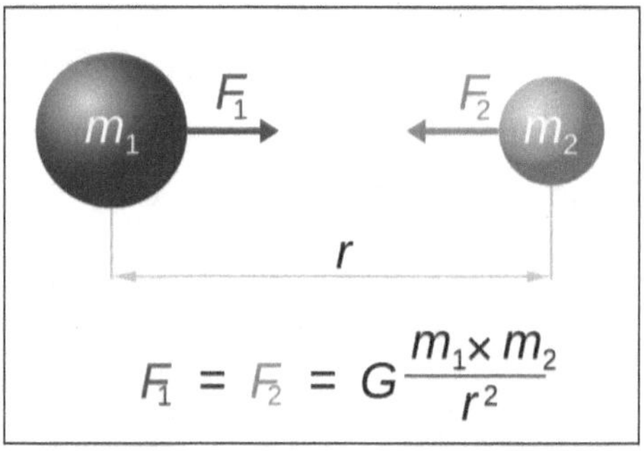

Figure 4. Gravitational force between two objects

We know that a force $F = G (m_1\ m_2)/r^2$ is acting on m_1 because of m_2, separated by distance r and G is the universal gravitational constant (Figure 4).

One of the participants then asked, what will happen to the force on m_2 if another object is brought close to m_2 and which is similar to m_2. This led to more questions coming up with more objects being added close to m_2. I then started discussing the difficulties in mathematics as we start adding more and more objects.

What happens if we have infinite objects like m_2, all lined up in a uniform pattern?

Now, infinite objects are easier to handle when we bring in the concepts of mathematics (infinite series).

One of my colleagues, Shri Vinayak Katdare, joined the following discussion and contributed to the mathematical calculations in the problem consisting of infinite series of balls.

This is an example of application of mathematics in a real physical problem. However, mathematics and physics application is a very natural integration; nevertheless the example is very interesting.

Suppose there is an infinite chain of balls of mass 1 kg each, kept at a distance of 1 m from each other, in a straight line. Now let us select a particular ball ZERO in the infinite series. Place a test ball T of mass 1 kg, at a distance of 1 m, perpendicular from the line of balls. Let us ignore the gravitational force due to Earth. The question is to calculate the gravitational force due to all the infinite balls on this one test ball T.

T

O

O O O O O O O O O O O O O O O O↓

ZERO

Figure 5. Infinite series of balls

Let the ball T be along Y axis and the infinite chain of balls be along the X axis with the central ball (numbered zero) at the origin.

Force due to the nearest ball (called as ball numbered zero) is $f_0 = G.1.1/1^2 = G$, directed along $\overline{-j}$ (vertically downwards in the direction of the arrow in the Fig. 5).

Forces due to all the other balls have X as well as Y components. All the X components on one side are cancelled by the corresponding components on the other side. Thus the resultant force is the resultant of all the Y components which are directed along $\overline{-j}$, negative Y axis.

Y component of the force due to the first ball (next to the nearest, i.e., next to ball number zero) is

$$f_1 = \frac{G.1.1}{(1^2+1^2)} \times \frac{1}{\sqrt{(1^2+1^2)}} = \frac{G}{(1+1^2)^{\frac{3}{2}}}$$

For the force due to the second ball (Y component),

$$f_2 = \frac{G.1.1}{(1^2+2^2)} \times \frac{1}{\sqrt{(1^2+2^2)}} = \frac{G}{(1+2^2)^{\frac{3}{2}}}$$

Similarly, due to the nth ball, the Y component of force is

$$f_n = \frac{G.1.1}{(1^2+n^2)} \times \frac{1}{\sqrt{(1^2+n^2)}} = \frac{G}{(1+n^2)^{\frac{3}{2}}}$$

Similar Y components will be there from the other side.

Resultant force due to all the balls (including ball number zero) is given by

$$f = G + 2G\sum_2^{\infty}\left(\frac{1}{1+n^2}\right)^{3/2}$$

∴ f=G(1+2 × 0.512435)=G(2.02487) =1.35 N, directed along $\overline{-j}$.

You must be aware that in the example gravitational acceleration has been used and hence it is situated in the vicinity of the earth. However, in infinite series the curvature of earth is ignored and so is the change of value of g if the balls are in a straight line. The idea here is more mathematical than physical to give the students an idea of how to use infinite series in physical situations.

04

Sea Water and its characteristics

The student who was curious on finding the velocity of water rising in the tree one day asked if this process can be used as a filter to get water out of the sea leaving salts behind, thus acting as desalination. This triggered a flurry of questions.

What is the salinity of the sea? Can it at all be reduced if all the pure water on earth is poured in to sea water? I had to assure them that by doing so the salinity will not change substantially.

Let us calculate the salinity of the sea water and then see if it is possible to use the concept of water rising in vegetation can be used for desalination. Is it even worth it? Is it economical?

The sea water is salty due to presence of several salts that make it difficult for humans to consume directly. Desalination currently is very expensive job and a lot of energy is required in conventional desalinations processes. However, these inefficient processes can be replaced with biological processes to convert salt water to potable water. Though pure water is trapped in the form of glaciers/ice all over the planet, it is important to preserve them as they are the "refrigerators" of the Earth and drive the water currents required for several phenomena related to temperature regulations. In order to understand the magnitude of the problems and the solutions available, it is essential to calculate the amount of water and ice available on Earth's surface.

By rough estimates, 70% of the surface of the Earth is covered by water and the remaining is occupied by landmass. The depth of sea/

ocean waters extends to as deep as 11 Km in the mid Pacific Oceans whereas at many other places they do not extend beyond a few hundred meters. So, one can assume, for simplicity, that the average bottom is flat at 2000 m. Also, for school level calculations it can be assumed that the bottom of the sea/oceans is flat.

The radius of the Earth is 6400 km which implies that its surface area is 5.15×10^{14} m^2. Assuming a uniform ocean depth of 2 km the volume of water (surface area multiplied by depth) is 1.03×10^{18} m^3. If 30% of the Earth's surface is landmass then the total volume of sea water is 0.721×10^{18} m^3.

Density of sea water varies across the surface of the Earth based on the fresh water inputs. But an average value of 1029 Kg/m^3 will give mass of water present in the oceans as 7.42×10^{20} kg (1.03×10^{18} m$^3 \times 0.7 \times 1029$ Kg/m^3). Surface area of the sea water is approximately 3.605×10^{14} m^2.

Currently there is about 30,000,000 km^3 of ice in the world's icecaps and glaciers. This volume of ice is fairly well measured (within 5-15%) by surveying the top of the icecaps with methods like radar and LASER altimetry, locating the bottom of the ice with methods like seismic soundings and calculating the difference. A breakdown is as follows.

The amount of ice present on the planet Earth is given in Table 1. This is an old table and the values keep changing as ice melts due to global warming. But we may assume these as some typical values.

According to the continental drift theory, the landmass of the earth was together. It can be assumed that the shoreline was circular in shape, again for the simplicity of the calculation.

Table 1. Volume of ice on surface of Earth

Location	Volume (km^3)	Fraction of world ice	Change in volume since 1960 (km^3)	Comment
Continental glaciers and ice fields	87,000 (± 10,000)	0.29 %	−4,700	grounded
Greenland ice cap	2,930,000 (2,620,000 to 3,000,000)	9.8 %	−2,000	grounded
Greenland continental glaciers	~50,000 (± 20,000?)	0.17 %	−350	grounded
Arctic Ocean pack ice	16,000 summer, 24,000 winter	0.01 %	−3,000	Floating
East Antarctic Ice Sheet	23,000,000 (21,800,000 to 26,040,000)	76.78 %	+10,000	grounded
West Antarctic Ice Sheet	3,000,000 (3,000,000 to 3,260,000)	10.0 %	−4,500	grounded
Antarctic Peninsula ice cap	227,000	0.76 %	(included with EAIS)	grounded
Antarctic continental glaciers	~50,000 (± 20,000)	0.17 %	−700	grounded
Ross Ice Shelf	230,000	0.77 %	−2,000	Floating
Ronne-Filcher ice shelves	344,000	1.17 %	−2,000	mostly floating
South polar pack ice	4,000 summer, 19,000 winter	0.08 %	+100	Floating
Total world ice	~30,000,000	100 %	−9,150	

Now consider the case if all the ice melts. How will it affect the sea level?

Total ice on the surface of the Earth is 3.00×10^{16} m^3. However, only the ice that is present on land will affect the level of the sea water. Ice that is floating will not increase the sea levels significantly when it melts. From the table, volume of water from ice on land is equal to 2.69×10^{16} m^3. This will spread out on all the ocean body having an approximate surface area of 3.605×10^{14} m^2 and thereby increasing the water level by 74.6 m. This figure is close the number of 70 m available in literature.

The content of salt water in seas varies depending on the local conditions; however, the average salinity of sea/ocean water is about 3.5% by mass.

A typical composition of sea water can be seen in Table 2.

Table 2. Composition of sea water

Seawater composition (by mass) (salinity = 3.5%)			
Element	**Per cent**	**Element**	**Per cent**
Oxygen	85.84	Sulphur	0.091
Hydrogen	10.82	Calcium	0.04
Chlorine	1.94	Potassium	0.04
Sodium	1.08	Bromine	0.0067
Magnesium	0.1292	Carbon	0.0028
Vanadium	1.5-3.3 mg/kg		

If one kg of sea water is evaporated the amount of chloride salts left as precipitates can be estimated from the data given in Table 2. Here, one has to assume that the more active elements will consume all the chlorine available and hence the calculation proceeds along the decreasing order of activeness of elements.

The atomic weights are H=1; O=16; Na=23; Cl=35.5; Ca=48; Mg=24; V=50; S=32; K=34; Br=79; and C=12.

Therefore, 1.08 g of Na is 0.047 moles of Na. One mole of Na requires one mole of Cl to form NaCl. Hence, 0.047 mole of Na requires 1.667 g of chlorine giving 2.747 g of NaCl.

Similarly, 0.04 g of K requires 0.0418 g of Cl giving 0.0818 g of KCl and 0.04 g of Ca requires 0.0592 g of Cl giving 0.0992 g of $CaCl_2$.

However, 0.1292 g of Mg is not consumed fully. Thus 0.0582 g of Mg is used to give 0.2303 g if $MgCl_2$, leaving 0.071 g of Mg available in other forms. We may assume that VCl is not formed with current assumption. Thus, salts which will form as precipitates are NaCl (2.747 g), KCl (0.082 g), $CaCl_2$ (0.099 g) and $MgCl_2$ (0.23 g). The total salts that will from precipitates amounts to around 3.5 g.

Taking the average salinity of seawater to be 3.5%, if all the ice melts, the addition of fresh water will reduce the salinity of the sea/ocean water. The salinity will change by total ice melting including which is floating.

The new volume of water will be $7.21 \times 10^{17} + (2.69 \times 10^{16}/0.9167) = 7.503 \times 10^{17}\ m^3$.

The new salinity will be $3.5 \times 7.21/7.53 = 3.35\%$

After the continental drift the shoreline can be considered to have increased by 7 times the unified landmass shore, for the 7 continents. What is the current length of shoreline available?

Surface area of the Earth is $5.15 \times 10^{14}\ m^2$. Since 30% of it is covered by land, in terms of flat geometry, it will be $1.54 \times 10^{14}\ m^2$. Assuming circular landmass, the radius for this circular geometry is 7.011×10^6 m. This gives a circumference of this circle as 4.406×10^7 m.

For 7 continents the shore line will be 3.084×10^8 m or 3.1×10^5 km

A research group studied growth of sea weeds XB and XG (not named yet) which can grow on in batches (for continuous availability of water) along the 1-km wide belt of the shore line. These sea weeds absorb water from the sea water leaving out all the salts. Thus, these sea weeds can then be crushed to obtain pure drinking water. It was observed that in 600 days, on an average 65 tons of XB and 37 tons of XG can grow in an area of 0.16 km^2 and 0.10 km^2, respectively. Since these sea weeds are mainly water, it was observed that 80% (by weight) water can be extracted for drinking purposes. If all the shorelines available to us can be used as a sea weed farm (1-km wide) to grow both varieties of sea weeds in this way, we can calculate the volume of pure water to be extracted.

The 1-km belt gives an area of 3.084×10^5 km^2. Since we get 102 tons of sea weeds from every 0.26 km^2 the total weight of the seaweed will be 1.187×10^6 tons. Since 80% of it is water, we would get 9.489×10^8 kg water for 600 days or 1.582×10^6 kg of water every day. This of course is a very small amount, but in future, better agricultural technologies can be used to enhance the yields. This example is an indication of the benefits of bio-technology to the society.

The data presented in the tables were from 2014 and they have changed over the years. The results therefore will also change.

We must also remember that sunlight is not the only source of heating of the Earth. It gets its heating (energy) from three different primary sources: Sun, Radioactive elements and heat trapped inside the core of the Earth.

We can estimate the amount of heat present inside the core of the Earth. It would be a rough estimate though, because there are no direct measurements of properties of the core. But from the figures that are known, let us attempt to see how this heat content can be estimated. Not that it is possible to get all that energy, at least not with the technology currently available, but we can always make an estimate. This would be explored in the next chapter.

05

Energy trapped in the core of the Earth

The core of the Earth is very hot and maintains a temperature of around 7000 K. This heat is trapped inside since the time of formation of the planet. The dimensions of the core are given in Figure 6, which shows how the temperature changes with the depth. This high temperature indicates the large quantity of energy trapped beneath the Earth's surface. Can this energy be tapped? If we used the energy at current rate of consumption, how long will it last? Of course this energy cannot be used up completely without understanding the consequences of the same on our environment. However, our aim here is to estimate the quantity of energy trapped under the surface of the Earth, which is called as the geothermal energy.

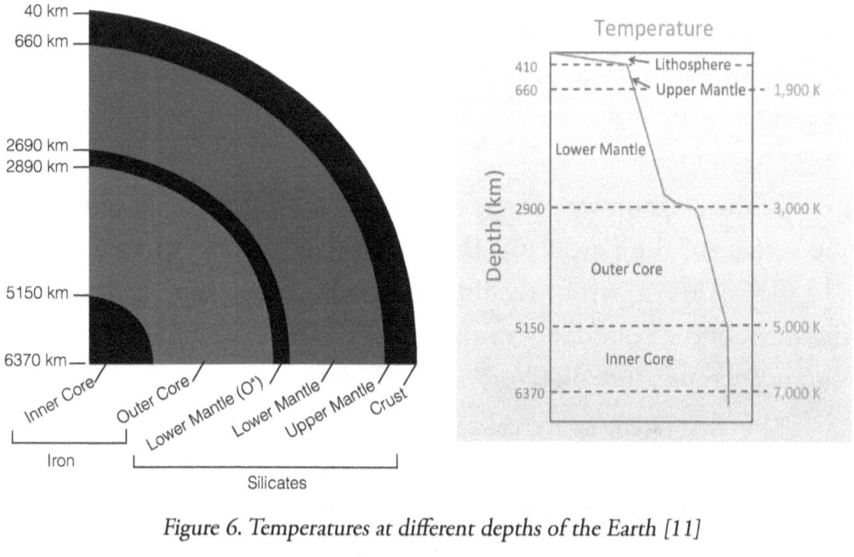

Figure 6. Temperatures at different depths of the Earth [11]

Heat flows constantly from its sources within the Earth to its surface. Total heat loss from the Earth is estimated to be 44.2 TW (4.42×10^{13} W). The Earth's crust effectively acts as a thick insulating blanket which must be pierced by fluid conduits (of magma, water or other) in order to release the heat underneath. More of the heat in the Earth is lost through plate tectonics, by mantle upwelling associated with mid-ocean ridges. Another major mode of heat loss is by conduction through lithosphere, the majority of which occurs in the oceans due to the crust there being much thinner and younger than under the continents' landmass (Figure 6).[11]

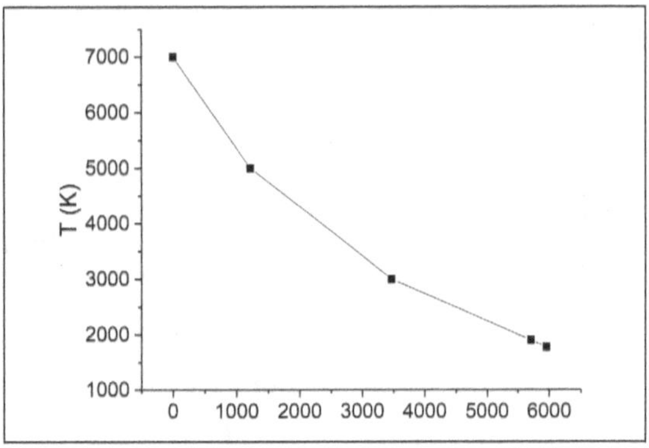

Figure 7: Temperature K (Y axis) vs. distance from centre of the Earth in Km (X axis)

The demand for energy in the world is increasing every day. Globally, the production of electricity in 2009 was 20,053 TWh. It has increased by 100% in last 8 years, and the current global electricity production is estimated to be 40,000 TWh. Geothermal energy can be one of the alternative sources of energy that can be used to minimise use of fossil fuels.

In the diagram we can see that the crust runs up to 40 km below the surface of the Earth, but the lithosphere extends up to 410 km. The crust is also a part of the lithosphere. Boundary temperatures and distances of the boundaries of shells from the surface are shown in the distance-temperature plot.

The first step is to calculate the temperature of Earth as a function of distance from its centre. This will give us an idea on how the temperature is changing as a function of distance. On the basis of current data, a linear change of temperature is assumed for simplicity.

It is assumed that the lower mantle data extends linearly up to the 2900 km. It is also desirable that average temperature of the 5 regions is plotted, assuming a linear change of temperature in each region.

The temperature at a depth of 2900 km is 3000 K. The temperature linearly falls up to 1900 K at a depth of 660 km. The linear gradient of the temperature is calculated by ΔT = (3000-1900)/(2900-660) = 0.49 K/km. Using this gradient, the temperature at the depth of 410 km can be calculated as $T = 1900–[(0.49 \times (660–410)] = 1777$ K.

The temperature at depth 0 km is 300 K. The gradient of this final shell is $\Delta T = (1777–300)/(410) = 3.6$ K/km.

The average temperatures are listed in Table 3 as a function of distance from the centre of the Earth.

Table 3. Average temperature for each shell

Distance from centre of Earth (km)	Average Temperature (K)
0 → 1220	6000 K
1220 → 3470	4000 K
3470 → 5710	2450 K
5710 → 5960	1839 K
5960 → 6370	1039 K

These are average temperatures for five different shells with their respective different temperature gradients. With five different shells having an average radius and an average temperature, as a simple approximation, one can estimate the heat content above the temperature of 300K (base temperature for calculations). Of course if the core temperature fell to 300 K, the surface temperature of earth which we call "room temperature" will not be 300 K. But to get a feel of the amount of heat content stored inside the surface of the Earth this approximation is justified. One can change the base temperature to a different value to get different estimates on heat content.

In order to calculate the heat content, let us assume that the 5 different shells have varying temperature but have same density of 5555 kg/m^3 and specific heat capacity of 1046.5 J/kg/K.

Table 4 gives the values of volumes of different shells and the heat content of each of them.

This gives a total heat content of 1.33×10^{31} J. Which is equivalent to $1.33 \times 10^{31} \times 2.8 \times 10^{-7}$ kW hour = 3.72×10^{24} kW hour = 3.17×10^{15} Tw hour. Even if we assume a 10% efficiency to convert this heat content into a productive electrical power, it amounts to 3.17×10^{14} Tw hour.

The total consumption/production of electricity today is around 40,000 TWh. Assuming the same rate of consumption, one may use geothermal energy for electricity production for 0.79×10^{10} hours, which is about 0.9 million years.

Table 4. Heat content of each shell

Shell No.	Inner radius (km)	Outer radius (km)	Shell volume (m^3)	Average temp (K)	Heat content (J)
1	0	1220	7.61×10^{18}	6000	2.5×10^{29}
2	1220	3470	1.67×10^{20}	4000	3.6×10^{30}
3	3470	5710	6.05×10^{20}	2388.5*	7.3×10^{30}
4	5710	5960	1.07×10^{20}	2388.5*	1.3×10^{30}
5	5960	6370	1.96×10^{20}	1038.5	8.4×10^{29}

* *This is the average temperature for combined shells 3 and 4 for making calculations simple.*

This is not the only pattern of heat trapped, inside the core of the Earth. In the areas of hot springs in India, the heat gradient is around 90 K/km[12] instead of 3.6 K/km that we assumed in the previous example. The figures derived are just a typical representation and that conditions in the last layer, i.e., crust of the Earth, can vary drastically depending on many factors. Also, getting all the trapped energy out is not so easy, simple and straight forward as mentioned in the example. But this is an attempt to get a "feel" of the quantity of energy trapped inside the Earth's surface. Moreover, a large amount of energy (equal to about 50 to 70% of the total amount) is also released inside the core due to radioactive decays, which has not been factored in the previous calculations.

Geothermal energy is one possible source of energy available for humans. There are other sources of energy, for example, sunlight. In the next chapter we will see the amount of energy we receive from the Sun, how it is distributed and how we can exploit it for our use.

06

Energy received from the Sun

Earth is situated at an average distance of 1.5×10^{11} meters from sun. The Sun releases 2.66×10^{41} photons per second for a wavelength of radiation 0.7 micro meters (700 nm). The radiation from sun is spread over different wavelengths of the electromagnetic spectrum. This intensity–wavelength curve can be seen in Figure 8. The numbers assumed here are "typical" and may not match with the Figure. These photons strike the surface of the Earth at uniform rate for which the Earth's atmosphere is transparent. When these photons strike a leaf of a tree, they enter the mesophyll cells of chloroplast. The photonic energy is used in the photosynthesis process, glucose is synthesized and Oxygen gas is released. For the synthetic process, CO_2 and H_2O are utilized. Considering only the wavelength

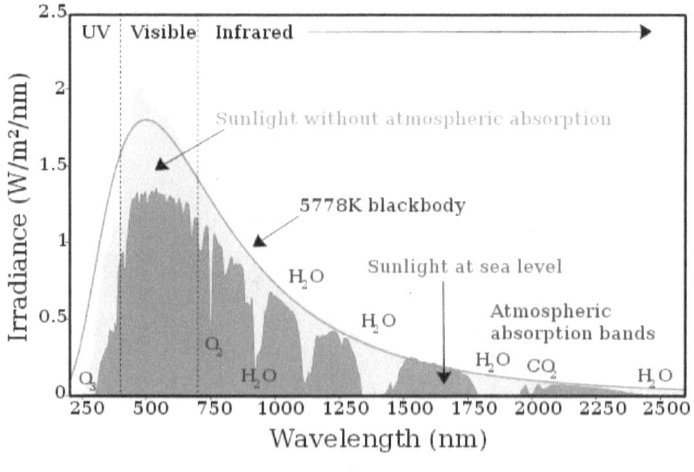

Figure 8. Intensity of sunlight received on surface of the Earth as a function of the wavelength. Source: https://www.wikiwand.com/en/Solar_irradiance

of 700 nm we can get a feel of how the energy is consumed. However, we should remember that radiations of other wavelengths are also effective and productive.

When these photons strike a leaf (having a flat surface area of 9 cm^2) normally, they enter the mesophyll cells of chloroplast. The photonic energy is used in the photosynthetic process, glucose is synthesized and O_2 is released. For the synthesis, CO_2 and H_2O are utilized. In order to keep the calculations simple, let us consider only the impact of 700 nm light. The same calculations can be extended to other wavelengths in an integrated way to understand the impact of all the visible light and also the wavelengths beyond the visible region of the spectrum.

As a first step it is essential to figure out the amount of energy that strikes the Earth per square meter of area.

The Sun emits 2.66×10^{41} photons/second at 0.7 micron wavelength. These photons are spread isotropically (equally in all the 4π direction around the Sun). Consider a sphere of radius 1.5×10^{11} meter. At the surface of this sphere all the photons will be equally distributed and the Earth is located (centre of the Earth) on the surface of the sphere. Since the photons emitted spread out in all the directions, the number of photons received on the surface of this sphere will be 2.66×10^{41} photons divided by the surface area ($4\pi \times 1.5 \times 10^{11} \times 1.5 \times 10^{11} = 2.83 \times 10^{23}$ m^2). Thus, the number of photons received by 1 m^2 on the Earth's surface is 9.41×10^{17}.

Let us assume the Earth to be a disc of radius 6.4×10^6 m to receive these photons and thus absorbing ($9.41 \times 10^{17} \times p \times 6.4 \times 10^6 \times 6.4 \times 10^6$) 1.2×10^{32} photons.

This amounts to the total energy deposited on the surface of the Earth as (hc/λ) 3.4×10^{13} J/s. This amounts to 2.67×10^{-1} J/s = 962 W (0.962 kWh) of energy received by area of 1 m^2 on the Earth surface. The actual value measured is 1000 W/m^2 after taking into consideration all the wavelengths and their attenuation by the Earth's atmosphere and other realistic conditions.

Since the area of a leaf is assume to be 9 cm^2 (9×10^{-4} m^2), the number of photons received by it (assuming it to be perpendicular to the incident photon direction) is 8.47×10^{14} per second. Of this energy, only 50% is absorbed by the leaf. On an average only 25% of the surface of the leaf is covered with chloroplast. Thus, only 1/8 of photons produce ionization, which means 1.05×10^{14} photons/sec. This is equivalent to 3.81×10^{17} photons/hour which are incident on this small leaf producing ionization.

$$6CO_2 + 12H_2O \rightarrow C_6H_{12}O_6 + 6O_2 + 6H_2O$$

For generation of 1 glucose molecule, generation of 5 electrons is required from chlorophylls. Since one photon generates one electron, there are 3.81×10^{17} electrons/hour. As 5 electrons are required to generate one molecule of glucose, $0.2 \times 3.81 \times 10^{17} = 7.62 \times 10^{16}$ reactions need to occur. This implies 7.62×10^{16} glucose molecules are generated per hour. This is equivalent to $7.62 \times 10^{16} \times 180/6.023 \times 10^{23}$ g of glucose, i.e., 22.8 μg of glucose per hour. For every reaction 6 molecules of oxygen are released and therefore 7.62×10^{16} glucose molecules will release $6 \times 7.62 \times 10^{16} \times 22.4/6.023 \times 10^{23}$liter/hour of oxygen. This is equivalent of 17 μl/hour of oxygen.

For every reaction 6 molecules of water are consumed. Thus, $7.62 \times 10^{16} \times 6 \times 18/6.023 \times 10^{23}$ g of water is consumed. It is 13.7 μg (nl) of water, in each leaf.

Now, let us compare this with the case when same amount of energy falls on a solar cell of similar physical dimensions.

If the leaf is replaced by a solar cell, it generates electricity at a potential difference of 0.6 V. With an efficiency of 10%, what is the amount of power generated by the solar cell per hour? On an average one photon knocks out one electron in the solar cell and the incident energy is not fully used to generate the output.

At 0.6 V there are 8.47×10^{14} photons striking a solar cell. Assuming each photon gives one electron and at an operation efficiency of 10 per cent we get 8.47×10^{14} electrons $\times$ 0.1 (eff) $\times$ 0.6 v $\times$ 1.6×10^{-19} or

8.13 μW of power. So, 8.13 μW of power per second equals to 29.3 μWh power generated by a solar cell having the same area as the leaf.

It is very interesting to see how chemical reactions and physics reactions can be compared in their output of energy by biological and physical systems. A very important point to remember will be that solar cells do not produce oxygen as a leaf does.

The oxygen released in atmosphere by the leaves is a slow process. Is it possible to release oxygen in some other process which is faster?

Theoretically it is possible. In the next chapter we will explore such theoretical models.

07

Emergency supply of oxygen

Space vehicles are launched with the air supply being maintained artificially inside the capsule, when outside the capsule the pressure is several orders of magnitude less (very high level of vacuum). The challenge is to device a mechanism to restore the provision for supplying breathing air in case of an accidental leak. Let us explore some innovative solutions here.

Let us assume, hypothetically, that ISRO launches a manned spacecraft to the Moon. The spacecraft, named JSO-12, consists of two identical sealed chambers separated by a wall. Two astronauts are travelling in one of the chambers where the life support system maintains an environment identical to the Earth's atmosphere. The other chamber is evacuated and contains various subsystems required for mission control, data handling and power generation.

Unfortunately, on the way to the Moon, the wall separating the two chambers develops a fissure and air from the astronauts' chamber leaks into the other chamber. The astronauts notice the warning signs on the life support system showing a sudden drop in pressure. They discover the leak in the wall and find that no more air is passing through the leak. This means that both chambers have attained equal pressure. They quickly fix the leak but the pressure in the chamber has reduced to half its original value. How can the air pressure in the chamber be restored? Is there an immediate solution to restore the pressure?

Let us assume that the volume of the astronauts' chamber is

50 m^3 and the initial temperature is 300K. The chamber occupying the equipment has the same volume. The volume occupied by the equipment is negligible. Assume that all gases behave ideally, and the composition of air is 80% nitrogen and 20% oxygen.

The air in the astronauts' chamber undergoes free expansion into chamber 2 (vacuum). Work done, W = 0 (since the gases are ideal) and since the chambers are thermally insulated, Q = 0.

$$Q = W + \Delta U$$

Therefore, the change in internal energy, $\Delta U = 0$

The temperature of the air remains constant, i.e., 300K.

Let P_i and V_i be the initial pressure and volume in the chamber. Similarly, P_f and V_f be the pressure and volume after the leak. Using Ideal Gas Law we can say $P_iV_i = P_fV_f$. This implies $P_f = P_i/2 = 0.5$ atm.

Now let us assume that the astronauts are carrying with them emergency capsules which release N_2 and O_2 rapidly when triggered. Capsule A contains 1 kg of NaN_3 along with some amount of KNO_3 and SiO_2, while capsule B contains 1 kg of potassium chlorate with catalytic amount of manganese dioxide powder. Will this work as an emergency mechanism? We have to calculate the quantities of the chemicals required. Appropriate safety measures are also to be taken when reactions are triggered using an electric impulse (heat) in capsule A and B. The following reactions take place:

$$2\ NaN_3(s) \rightarrow 2\ Na + 3\ N_2\ (g)$$

$$10\ Na + 2\ KNO_3 \rightarrow K_2O + 5\ Na_2O + N_2\ (g)$$

K_2O and Na_2O released in these reactions react further with SiO_2 to give stable solid products. No sodium is left unreacted after these two reactions.

In capsule B we get $2\ KClO_3$ (solid) $\rightarrow$ $2\ KCl$ (solid) + $3\ O_2$ (gas)

The next logical step would be to calculate the quantity of material in capsule A and B.

The volume of air required to be replaced is half of the volume of the chamber (which is 50 m^3) and will be 25,000 l. Thus one requires 20000 l of Nitrogen gas and 5000 l of oxygen gas.

One mole of oxygen, assuming all at NTP, is 22.4 l, and 2 moles of $KClO_3$ delivers (22.4 × 3) 67 l of O_2, when heated in presence of MnO_2.

The requirement is (5000/67) × 2 moles = 14 moles of $KClO_3$, which amounts to 1.7 kg of potassium chlorate.

But requirement of Nitrogen is quite heavy. Here, 22.4 l of N_2 gas is given by 10 moles of NaN_3. Thus (20000/22.4) × 10 = 8928 moles of NaN_3, which is equal to approximately 580 kg of NaN_3, would be needed for obtaining this amount of Nitrogen. This is the same reaction which is used in some of the safety airbags. But carrying 580 capsules of 1 kg each is quite difficult for a spacecraft and hence we must think of some alternatives.

If they fail to restore the original pressure and temperature condition and are forced to remain there for a prolonged period, it can impact their health. Their Systolic and Diastolic blood pressure will increase. The rate of respiration will also increase and their ear drums can burst too.

We must find more effective solutions along these lines. It is also worthy to note that the cost of carrying 1 kg of material in a rocket is roughly around USD 20,000. This high cost is due to the additional amount of fuel needed at the initial stage. This additional fuel in turn increases the total weight of the rocket and so on. It is a classic case of 'positive feedback loop' in electronics that we would learn in higher classes.

But why is it so costly to take things up in the rocket? In the next chapter we would see how a jet engine attached to a sled works in a rocket.

08

Chemistry and physics of a jet engine

In this chapter we will see how to calculate the dynamics of a sled which is run by a jet engine. Consider a jet engine in which the combination of the fuel and oxygen is burnt in the combustion chamber. This combustion generates hot gas which is released at a high velocity providing the required force for the motion of the vehicle fitted with this engine. Let us assume that a 20-kg sled is fitted with a jet engine that burns methane as a fuel. It has a maximum fuel burning rate of 12 g/min. The chemical reaction is

$$CH_4 + 2O_2 \rightarrow CO_2 + 2H_2O + 890 \text{ kJ/mol}$$

Let us assume that the jet engine has 50% efficiency, which means 50% of the energy generated by burning the fuel is converted into mechanical energy. When the burning rate is maximum, the jet provides sufficient force for the sled to accelerate at a speed of 30 km/h in one minute along a straight line over a plane ice-covered ground.

Let us calculate the rate of exhaust from the engine. From the previous equation we can say 16 g of fuel (CH_4) generates (along with 64 g of O_2) 80 g of exhaust (CO_2 + $2H_2O$). Hence 12 g of fuel will generate 60 g/min of exhaust.

This means mass of fuel will generate mass of exhaust gas. The exhaust gas molecules all pick up different velocities depending on the dynamics of the fuel burning inside the chamber. However, the distribution does lead to an average velocity of the gas molecules.

Molar mass of $CO_2 + 2H_2O$ is 80. Thus, 60 g (mass m) of exhaust is 60/80 = 0.75 mole. Energy release is 890 kJ/mole. Therefore, for 0.75 mole the energy released will be 0.75×890000 J.

Half (50%) of this energy is utilized as kinetic energy. So, $1/2\ mv^2 = 0.5 \times 0.75 \times 890000$ J.

m = 60 g = 0.06 kg. Therefore, $v = \sqrt{\frac{0.5 \times 2 \times 0.75 \times 890000}{0.06}}$

=3335.0 m/s

Once we know the average velocity of the molecules and the number of molecules being ejected it is easy to calculate the thrust provided by the jet engine.

Mass rate of exhaust = 60 g/min = 1 g/s = 0.001 kg/s = dm/dt

Average exhaust speed = 3335.0 m/s = v

Force = rate of change of momentum = (mass rate of exhaust) (average velocity of exhaust)

= (dm/dt) × (v) = 0.001× 3337.5 N = 3.3375 N

In spite of the large thrust provided by the jet engine, the sled moves with an acceleration which is lower than what it should as a result of this thrust. Thus the resistive force experienced by the sled can be easily calculated.

From rest ($u_s = 0$), the sled attended a velocity v_s = 30 km/hr = 75/9 m/s in t = 1 min = 60 s

Average acceleration of the sled $a_s = \frac{\frac{75}{9} - 0}{60} = \frac{5}{36} =$

Mass of the sled, M_s = 20 kg. Average 'accelerating' force, $f = M_s.a_s$ = 25/9 N = 2.78 N

However, the reaction force experienced by the sled is 3.34 N. Thus the average resistive force = 3.34 – 2.78 = 0.56 N

This resistive force is required to be balanced by burning fuel at a constant rate to run the sled at a constant velocity. This constant rate of burning of fuel can be calculated easily on the basis of the resistive force.

The power required for maintaining the velocity = Average resistive force × (constant) average velocity = 0.56 × 75/9 W = 4.67 W.

For producing 890000 watts, the rate of fuel burnt will be dM/dt = 16 g/s.

Thus for 4.67 W, $dM/dt = \frac{4.67 \times 16}{890000} g/s = \frac{4.67 \times 16 \times 60}{890000}$ g/min
= 0.005 g/min (5 mg/min)

This is an attempt to see how principles of chemistry and physics can be integrated to understand dynamics of a sled which is run by methane gas. Further dynamics can be added as per our interest.

Now, for producing 890 kW of power, 16 g of fuel needs to be burnt, and for 4.4 W, 7.9×10^{-5} g of fuel needs to be burnt per second which is 4.7 mg of fuel per minute, as compared to 12 g per minute during the acceleration phase.

At this point we can see that the case is quite similar in case of air flights, where most of the fuel consumption happens during the acceleration and take off. Once the aircraft has reached its cruising altitude, there is very little fuel consumption and can be a case study for further investigation.

If we change the rocket vertical now you will see how it has to act against the gravitational force and hence the situation gets more complicated. This is the prime reason why it costs so heavily to take 1 kg into the space. Space agencies work very hard to reduce every gram of weight that they can from the satellite's design and these optimizations are special fields in engineering.

During another tea session, looking at the amount of Kcal written on the packet of a potato chip, we entered into a very interesting discussion on how much of the energy we consume actually gets converted into kinetic energy in our body.

In the next chapter we will explore how energy conversions occur in living systems with the example of a cheetah.

09

Dynamics of a Cheetah's run

Cheetah (scientific name *Acinonyx jubatus*) is the fastest land animal. Now extinct in India, it is found mostly in Africa and Iran. When in its natural habitat, it can accelerate from rest to its maximum running speed of about 30 m/s (about 100 km/hour) in just 3.0 s. This acceleration of around 10 m/s^2 is nearly equal to the acceleration due to gravity, amounts to a most impressive acceleration. By comparison, consider that a fast sports car - a Porsche - takes about 4.0 s to attain the same speed.[13] Speed of 30 m/s is Cheetah's main weapon for hunting where it can swoop down on preys very swiftly, and at the same time, it is also its defence from predators. But the cheetah must put in great effort to attain this speed and move its body machinery with unusually high efficiency, consuming lots of energy. At the same time, it must manage the high amount of fatigue thus generated. This kind of high speed is not sustainable for a very long time and the cheetah cannot run at this speed for distances longer than 500 m. The longest unsuccessful chase at very high speed has been recorded to last a distance covering 650 m.

The head-to-tail (base) length of a typical cheetah ranges from 110 cm to 150 cm (43 to 59 in). The length of the tail can measure from 60 to 84 cm. [13, 14, 15] Their front legs are around 66 to 94 cm long. Compared to a leopard, Cheetah are longer, taller and have longer tails. At high speeds, a Cheetah breathes about 60 to 150 times a minute. A cheetah weighs between 45 to 60 kg. In order to understand some of the other attributes of cheetah we assume its mass to be 50 kg is assumed.

Figure 9: A Cheetah running at a high speed [16]

A cheetah reaches its maximum velocity in around 3 seconds and can run at its maximum speed for around 20 or 21 seconds before it needs to slow down. In this duration its body temperature rises by around 1.5 degrees C. Any further rise in body temperature can turn fatal for the animal. In this duration a Cheetah requires more than 50 l of oxygen but can often manage with a smaller quantity. A Cheetah can identify its prey at a distance of around 5 km.

When running at a high speed, a cheetah takes only about 220 milliseconds to swing its legs back to the front after having finished a stride. In each stride a Cheetah jumps around 7 m. This is the limiting factor on how fast it can run.

It is very interesting to understand the dynamics of a cheetah as it runs to capture its prey. Students can understand how a Cheetah gets its energy from different manifestations of respiration. The example of a Cheetah also highlights how Nature functions at the extreme and how speed, which is a weapon for the Cheetah, is also its drawback and limiting factor. The calculations below have several assumptions by which a student of secondary school level can determine the numbers with ease.

We have already seen that the acceleration of cheetah is 10 m/s^2 and it can run at a speed of 30 m/s for 21 s. The distance covered during this can be calculated in two phases. Distance travelled during the acceleration phase $d_1 = 1/2\ at^2$ or $d_1 = 1/2 \times 10 \times 3^2 m = 45m$.

In the second phase it runs at 30 m/s for 20 s which amounts to 600 m. Thus, total running distance is 645 m.

The cheetah has to do work against friction, mostly due to air. The other resistive forces are also effective but may not be included in the calculation. Assume that this frictional force is always 100 N. We can calculate the total mechanical work done by the cheetah during the first 23.0 s of its motion.

The work done can be calculated in many ways. One way is to calculate the change in kinetic energy and overcoming the frictional force over the distance the cheetah covers in 23 s (645 m).

Work done in change in kinetic energy $K=1/2\ mv^2 - 0$, which is

$$K=1/2 \times 50 \times 30^2\ kJ=22.5\ kJ$$

Overcoming friction due to air $W_d=F_a \times (d + d_1)$,

Total work done $W = K + W_d$ which is $W_d = 100 \times (600 + 45) =$ 64.5 kJ

During this run, both the mechanical work and the heat produced by the metabolism in the body are significant. It is estimated that the body temperature rises from 38.5^{0}C to 40.0^{0}C during this 23-seconds run. If we assume that this rise in temperature is linear in time, we can calculate the heat generated from the metabolism.

Of course the Cheetah is connected to the environment and hence there will also be heat loss in these 23 seconds. But neglecting this heat loss, the total energy required to raise this temperature can be calculated simply. Further, it is justified to assume that like all other vertebrates, Cheetah is mainly constituted of water, and hence for the sake of calculations, we can assume it to be water of mass 50 kg with specific heat capacity of 4.2 kJ/kg/K.

Heat generated in the body of the cheetah, $H = Sm\Delta t$

$H = 4.2 \times 50 \times (40\text{–}38.5)$ kJ = 315 kJ

Thus the metabolism generates energy, from which a fraction of the energy raises the temperature of the body and the remaining is used for the mechanical work. Therefore, total energy generated in the body of cheetah, E = H + WE = 315 + 87 kJ = 402 kJ. Thus the fraction of energy used up in raising the body temperature is Fraction= 315/402 = 0.78

Out of the total amount of energy generated during metabolism only 6% is converted to kinetic energy. Rest is utilized in overcoming the resistive forces during the run at a high speed.

This poses the next obvious question, where does it get all the energy from. From "physics" the question then becomes one involving "chemistry", – the concept of integration which we discussed in the earlier chapter. "Biology" of the question will come later!

When the cheetah starts running, it generates its energy initially by aerobic respiration, in which glucose is oxidized in the presence of oxygen, resulting in generation of ATP. In this process, each mole of glucose generates 36 moles of ATP, and 1130 kJ energy is released when all these ATP molecules are utilized. Running at high speeds increases the demand for oxygen, resulting in the increased breathing rate of 150 breaths per minute. The balanced reaction of the process is given by

$$C_6H_{12}O_6 + 6O_2 \rightarrow 6CO_2 + 6H_2O$$

What will happen if the requirement of all the 402 kJ of required energy is to be obtained by aerobic respiration? Let the molar volume of Oxygen be 22.4 l.

Number of glucose molecules required to produce 402 kJ of energy is

n_g = (402 kJ)/(1130 kJ) = 0.355

Number of moles of O_2 is $n_O = 6n_g$

$n_O = 6 \times 0.355 = 2.1$ moles

Volume of O_2 required, $V=22.4 \times n_O=22.4\times2.1$ l=48 l

This Oxygen is extracted from the air while breathing. The inhaled air (about 500 ml per breath) contains 20.0% of oxygen by volume. The exhaled air is assumed to contain 15% (by volume) of oxygen. Thus the total oxygen that the cheetah breathes in 23 seconds, at the rate of 150 breaths per minute is,

Amount of O_2 absorbed by the lungs per breath V_1 =5% of 500 ml=25ml is

Total number of breaths during its entire motion, $n=N/60 \times$ total duration of motion $n=150/60\times23=57.5\approx58$

Thus the amount of intake of O_2, $V_a=n\times V_1=58\times0.025$ l=1.45 l

The total volume of oxygen it can breathe is 1.45 l as compared to its requirement of 48 l even at this high rate of breathing. The volume of the lungs also plays an important role in deciding this quantity. Thus it is clear that the energy requirement of the cheetah's muscles is not met by the aerobic respiration. From the biology textbooks, we know that under such circumstances, it is the anaerobic respiration that generates ATP molecules. But, here only 2 molecules of ATP (instead of 32) are generated per molecule of glucose.

Anaerobic respiration converts the energy from glucose into ATP where if one mole of glucose were used (burnt), the energy released is 2872 kJ.

Energy generated in production of one mole of ATP due to oxidation of glucose is $U_g=1130/36=31.4$ kJ

Whereas in the anaerobic respiration it is 2872/2 = 1436 kJ. Thus the efficiency factor is 31.4/1436= 0.022

Hence the amount of glucose required if all the energy is obtained, and which in this case is from anaerobic respiration, can be calculated as follows.

Amount of glucose of required, $N=402/2U_g$

$N=402/62.8=6.4$ mole

Approximate molecular mass of glucose,
$M=6\times12+12\times1+6\times16=180g$

$m=180\times6.4=1.2$ kg

The acceleration of a cheetah is almost same as that of a body falling under the influence of gravity on the surface of earth. At such high value of acceleration and the velocity at which it runs, of the total energy consumed, only 6 per cent goes in changing its kinetic energy. If it was to breathe normally during 23 seconds, a cheetah would need 50 l of Oxygen, which is a huge amount considering the fact that it weighs around 40 to 60 kg. However, given the respiratory structure of a cheetah, it cannot breathe more than 1.5 l of oxygen in this duration. Thus switching over to anaerobic respiration it consumes nearly 1 kg of glucose to achieve this feat in about 20-25 seconds of running. After calculating these numbers, we can well imagine the extreme conditions that a cheetah can routinely set itself up against, for short durations, in order to achieve its kill.

The calculations on the basis of several simplifications give us an idea on how a cheetah consumes food (quantity in terms of glucose) and converts it into the energy required to run at high accelerations and high speeds. This question was part of the International Junior Science Olympiad 2013, held at Pune.

The author would like to thank B. Belapurkar, P. Burma, M.M. Chaturvedi, A. Dighe, P. Ghalsasi, U. Ladiwala, S. Narwekar, A.A. Natu, S. Raychaudhuri, and D. Mathur for their contributions at different stages. This can also be seen at reference https://www.tifr.res.in/~pkjoshi/articles/Cheetah_DM.pdf

The discussion we had was inspired by a movie on Cheetah, produced by the National Geographic, which is available on YouTube. (https://www.youtube.com/watch?v=B4nd9GF1dRg or https://vimeo.com/53914149) You may watch it to understand a few other features of the animal. Cheetah runs with its four legs moving independently and hence these four legs are in air or touch the ground at different times. However, you can see the synchronization of muscular activities (sometimes called as coordination in Biology) which leads to its head moving almost along a horizontal line allowing a great focus on its prey while running at high speeds.

Also, the time it takes for any leg to leave the ground and come back touching again is around 0.22 s. In this duration the cheetah travels around 7m, resulting in a upper limit of speed of $7/0.22 \approx 30$m/s.

Also, cheetah can resolve a 1 m object at a distance of around 5 km.

You should now get a fairly good idea on how science is all combined and interconnected, when it comes to nature and there is no "compartments" between physics, chemistry and biology.

You may also think that how the head of the cheetah remains stationary in spite of the four legs changing their positions so rapidly. Also, how can a cheetah have such a large resolving power?

In the next chapter let us explore the example of a large trailer truck to see how forces are balanced. It is easy and interesting at the same time to study a physics example and get a feeling of how it is applied in biological systems.

10

Balancing forces on a vehicle

We all have seen large trailer trucks carrying heavy load so that the load is distributed on the wheels. It is interesting to see how the load is shared by each tyre. The problem is simplified by considering some parameters which make it easy to calculate the required numbers.

Let us assume a situation where one such trailer truck has 20 wheels. Assume for simplicity that there are 10 wheels on both sides of trailer and they are in a row of single wheels right from the beginning to the end of the length of the trailer. Each wheel has a mass of 100 kg. Also assume that the platform is rigid and remains flat and horizontal when the load is placed on the platform (Fig. 10) as shown in the representative figure.[17]

Figure 10. A representative figure of heavy load

When the load is placed on the platform, the platform moves downwards by 15 cm. Assume that all the tyres are supporting the platform by uniform springs with spring constant K = 10000 N/m. After placing the load, the trailer began to move on a smooth road that is free from any jerk. At one point it encountered a pothole which was 30 cm lower than the road level and it was long enough to accommodate two tyres (on the left side) simultaneously.

Can we find the difference/gap between the bottom of the tyre and the surface of the road inside the pothole? The following discussion will give us an idea to imagine the real-life situation and take all the assumptions into account for making such calculations.

As assumed earlier, when the load is placed on the trailer, the platform is displaced by 15 cm. Since there are 20 springs which displace by 15 cm, the total force on the platform is

$F=20 \times 10000 \times 0.15 \text{ N} = 30000 \text{ N}$

Assuming acceleration due to gravity as 10 m/s^2, the load placed on the platform is 3000 kg.

Now when the two tyres are suspended in the pothole the load shifts to the remaining 18 tyres. However, we will have to keep track of the two tyres which are now suspended from the platform and thus added to the load.

This results in total load as 3200 kg, leading to a total force on the platform of 32000 N.

The load is now distributed on 18 tyres (though the distribution is asymmetric).

So, the displacement of the spring or the displacement of the platform from the initial position of no-load situation will be X_1= 32000/(18×10000)=17.77 cm.

Thus when the two tyres entered the pothole the platform was displaced by an additional 2.77 cm.

At this time there is no load on these two tyres and hence the springs restore to its original length thus gaining 15 cm.

But the tyres weigh 100 kg each and thus they cause additional spring extension. This extension X_2 = (100 × g)/k=(100 × 10)/10000 = 0.1 m = 10 cm.

This implies that the total extension of spring is 15 cm + 10 cm = 25 cm. We have to add additional 2.77 cm lowering of the platform, from where the extended spring supports the tyre, when the load is redistributed among 18 tyres.

Hence, the bottom of the tyre is lowered by 27.8 cm from the road level on which it was traveling, i.e., 2.2 cm above the bottom of the pothole.

This discussion is based on a question in IJSO camp. The question was designed to test the ability of the students to explore all the possibilities of a practical situation. If we do not consider the additional extension of the springs when the two tyres are suspended in the pothole, we will get the calculations wrong.

The question/situation described earlier, gets more complicated if the depth of the pothole would have been less than 27.8 cm because then the spring will not be fully extended and the load will not be zero from these two tyres. Thus, the first assumption of redistribution of load on 18 tyres will not be reasonable and the calculations will proceed in a different way.

As we promised to have some more discussions related to biology, let us look at how a convex lens system works before we try to appreciate the power of eyes of cheetah.

11

Convex lens and its applications

Bifocal convex lens is always included at secondary school level. Understanding the construction and working of bifocal convex lens is very essential in physics and biology-related applications and can be approached in very many interesting ways. The dynamics of the convex lens can be written by the equation $1/u+1/v = 1/f$. However, as soon as this equation is displayed, the students (and teachers also!) would say that this equation is incorrect And the correct equation would be $1/v - 1/u = 1/f$. However, this is the equation when the lens is placed at the origin, i.e., $X = 0$, the object is placed along the negative

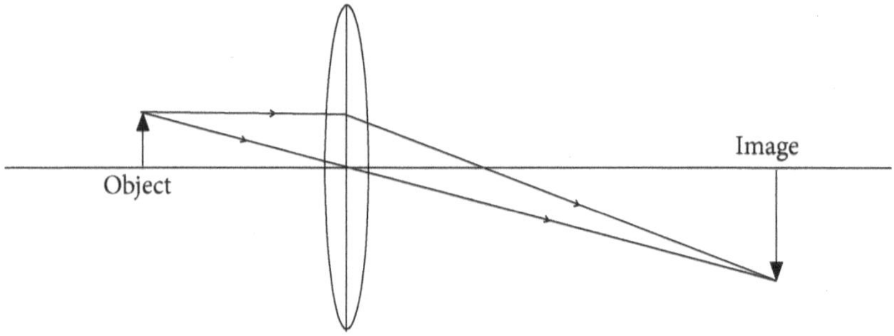

Figure 11. Ray diagram of a convex lens. The bending of light at center of the lens is only representative. Light bends at both surfaces of the lens.

X axis and image is formed along the positive X axis. The focal point is also along the positive X axis. Thus the distance between the lens and the object will be $-u$, the distance between lens and image will be v and the focal length of the lens is f.

If one puts –u in the equation above, it turns into the first equation. There is nothing wrong with the second equation, but while conducting the experiment, one has to keep track of the coordinate system. Thus many textbooks refer to the first equation as a "working equation".

There are many advantages of using the "working equation". First, we do not have to worry about the coordinate system. Second, it gives a physical feeling of the real situation, i.e., addition of 1/u and 1/v is a constant number and it is equal to the inverse of focal length. Indicating that as 1/u increases (decreases) 1/v decreases (increases). On a few occasions, while conducting the experiment students use the textbook equation, i.e., the second equation, but forget to keep track

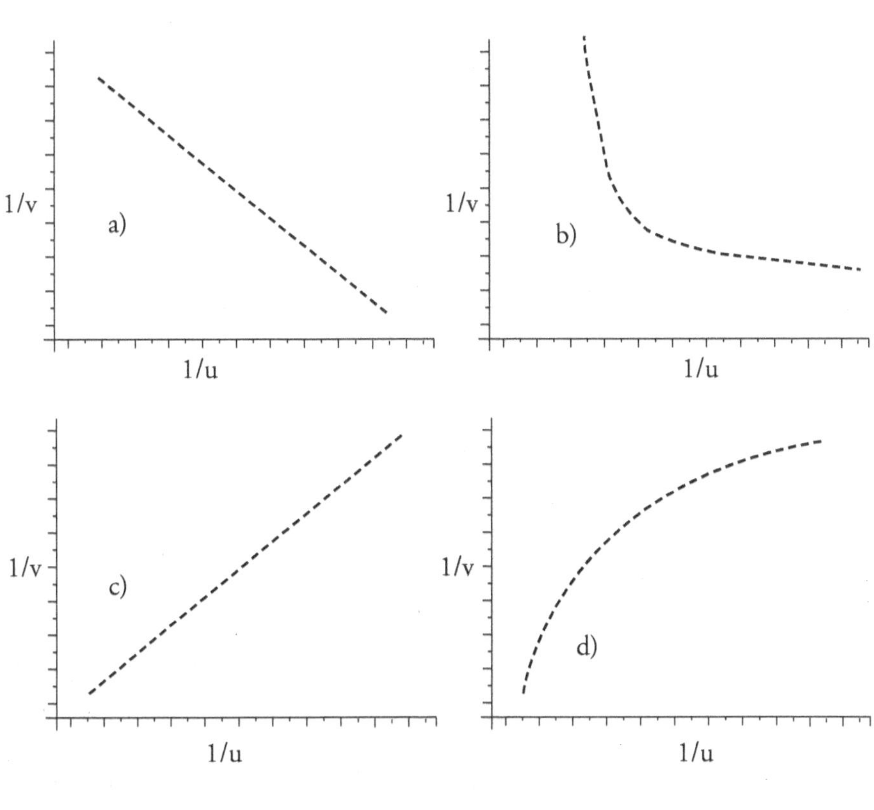

Figure 12. The relationship between 1/v and 1/u. The graphs are not to scale.

of the coordinate system resulting in a physical incorrect solution. The resultant graph of 1/v vs. 1/u they get is therefore wrong.

When the students were asked to choose the correct relationship, Figure 12, that represents a convex lens, more than two-thirds of the hundred students gave the wrong option. After discussing the conditions of the u, v and f, most of them arrived at correct option 'a'. Choosing the correct option becomes simple for those who have conducted the experiment.

For those who have not conducted the experiment, the solution comes by making 3 most obvious assumptions. The first one is object being at infinity, $u = \infty$ leading to $v = f$. Similarly, $u = f$ leads to the condition of $v = \infty$. And finally the third condition is $u = v = 2f$. Thus, it leads to a straight line. This explains why option 'a' is the correct one.

Now the situation can be made more complicated by asking the relationship between u and v instead of 1/u and 1/v.

In an experiment it is u and v which are measured and hence understanding the relationship between u and v gives us a better visualization effect of the reality with convex lens.

It is now observed that only few students could give correct answer to this question. The easiest way reach the right answer would be to ask the same question: identify the three most obvious conditions.

Here, the first is with $v = \infty$, i.e., the image is formed at infinity when the object is placed at the focal position, $u = f$ and so on. Now you can easily arrive at option 'b' as the correct answer in figure 12.

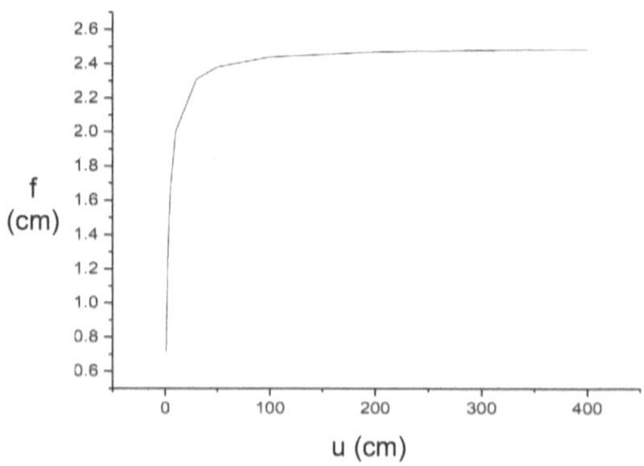

Figure 13 Relationship between f and u in eye

The next one would be to describe the most common phenomenon/

application where this concept of u, v and f as in convex lens is applicable. As stated earlier, the most common answers were spectacles, microscope, telescope etc. However, after lots of prodding we arrived at the answer - 'eyes'. This is because most of the time we perceive 'lens' as part of 'physics' but 'eyes' as part of 'biology'. It is extremely important to introduce science as an integrated study rather than treating it as compartmentalised as physics, chemistry and biology.

Once we arrived at the answer of eyes, the next obvious question was "what is so unique about eyes?" in terms of u, v and f. The answer is that here v = constant (almost) for eyes. Thus f has to change with changing object-lens distance, u. The variation of f with u is given in Figure 13.

From the graph it can be seen that as object-lens distance decreases, the focal length reduces only marginally remaining practically constant. But after the object-lens distance u is comparable to v (the diameter of the retina), the focal length begins to reduce at a much faster rate and below the size of the retina, the focal length reduces at an extremely fast rate.

This explains the stress one feels when an object is placed very close to the forehead between the eyes, in which case when the muscles are stressed to reduce the focal length to its extremes.

This does not however explain how the cheetah can have such high resolving power to observe its target/prey. But we must remember that cheetah is not a very powerful animal, and hence its weapons are early warning (resolving objects at large distances) and its great speed. This combination can have many applications in human life and you may explore them to get some amazing insight.

References

1. S. Ruben, M. Randall, M. Kamen and J.L. Hyde. Heavy Oxygen (O^{18}) as a tracer in the study of photosynthesis, J.Am. Chem. Soc, 63(3), 1941, p. 877-879.
2. https://www.thoughtco.com/how-much-oxygen-does-one-tree-produce-606785 [Accessed on 19 November 2020]
3. R.A. Osborne and R.J. Geider. The minimum photon requirement for photosynthesis, New Phytol (1987) 106, p. 631-644.
4. http://hyperphysics.phy-astr.gsu.edu/hbase/Biology/ligabs.html [Accessed on 19 November 2020]
5. https://www.itslife.in/gardening/trees/neem-tree [Accessed on 19 November 2020]
6. Fleur Longuetaud et.al. Picea abies sapwood width: Variations within and between trees, Scandinavian Journal of Forest Research, 2006; 21: 41
7. M.E. Olson and J.A. Rosell. Vessel diameter – stem diameter scaling across woody angiosperms and the ecologiclal causes of xylem vessel diameter variation, New Phytologist (2013) 197, p 1204-1213.
8. Dana Rademaker. Arabidopsis thaliana under far-red light. https://esc.fnwi.uva.nl/thesis/centraal/files/f1513550460.pdf, p7 [Accessed on 19 November 2020]
9. http://plantphys.info/plant_physiology/photoeco.shtml [Accessed on 19 November 2020]

10. https://www.fs.fed.us/global/iitf/pubs/sm_iitf070-8.pdf [Accessed on 19 November 2020]
11. https://malagabay.wordpress.com/2017/06/22/flipping-geology-walking-on-sunshine/geothermal-gradient/ [Accessed on 19 November 2020]
12. H. Chaudhuri, C. Maji, K. Seal, S. Pal, and M.K. Mandal. Exploration of geothermal activity using time series analysis of subsurface gases data from Bakreshwar hot springs area, eastern India. Arabian Journal of Geosciences (2018) 11; p. 324.
13. "Cheetah: fast facts", Zoological Society of London. Accessed on 5 June 2013
14. L.K. Boast et al. Regional variation in body size of the cheetah (Acinoyx Jubatus) Journal of Mammalogy, 94(6); (2013) p. 1293-1297.
15. https://animalrespriation.weebly.com/cheetah.html [Accessed on 19 November 2020]
16. https://www.cat-lovers-only.com/cheetah-speed.html [Accessed on 19 November 2020]
17. http://www.wermac.org/rigging/lifting_rigging_part1.html [Accessed on 19 November 2020]

Garuda Prakashan Books

Garuda Prakashan Books

THE GARUDA CLUB

BE THE CHANGE

GET OUR BEST BOOKS EVERY MONTH

at a 40% DISCOUNT

GRPR.IN/GC

- Our best Book shipped free every month
- Best-selling Authors
- Game changing Scholarship

Scan Me

*T&C apply.

गरुड

Register:

Please register your book purchase at **grpr.in/register** to stay in touch and get informed about upcoming books!

To order:

www.garudabooks.com

Follow us:

Website: **www.garudabooks.com**

www.facebook.com/garudaprakashan/

@garudaprakashan

@garudabooks

/garudabooks

Contact:

Email: contact@garudabooks.com

International queries:

Email: international@garudabooks.com

www.ingramcontent.com/pod-product-compliance
Lightning Source LLC
LaVergne TN
LVHW090335210926
848480LV00004B/1416